Windows® 11 Intr

Quick Reference Guide

Comprehensive video training & instruction man ...mp.com

TEACHUCOMP, INC.®
...it's all about you

Windows Basics

Understanding the Taskbar

1. Widgets (News/Weather/Etc.)
2. Start Button
3. Search
4. Task View
5. Teams
6. Pinned Apps
7. Hidden Icon Menu
8. Hardware Status Indicators
9. Date/Time & Notification Center

The Start Button

The "Start" button is located in the middle of the Taskbar at the bottom of the Windows screen, by default. When activated, it opens the "Start Menu," providing access to apps, files, and folders. In this guide, we will refer to this button as "Start."

Starting Windows

1. Click or swipe up on the starting picture.
2. Select a user, if necessary.
3. Sign-in with the method last used. ***Or***, click the "Sign-in options" link to select a different method.
4. Sign-in methods vary by hardware configuration:

Password: Type your password and press the "Enter" key.

PIN: Type your stored PIN password.

Facial Recognition: Look in camera for Windows 11 to recognize your face or iris.

Fingerprint: Place finger onto installed fingerprint reader to sign in.

Security Key: Third-party hardware usually in the form of a USB or NFC device.

Signing Out and Shutting Down

1. **To lock the screen**, click "Start," the "Power" button, and then "Lock." ***Or***, press the "Windows" and "L" keys. +L
2. **To unlock the screen**, swipe up or click anywhere on the screen and sign in.
3. **To sign out**, click "Start," your account picture, then "Sign out."
4. **To sleep, shut down, or restart**, click "Start," then the "Power" button, then the desired option.

The Start Menu and the Taskbar

Understanding the Start Menu

The Start Menu provides access to apps, files, and folders. **To access the Start Menu**, click "Start." ***Or***, press the "Windows" key on your keyboard. Once activated, the Start Menu displays a search bar at the top, a list of apps "pinned" to the Start Menu, a list of previously opened ("Recommended") files, and the "Account" and "Power" buttons at the bottom:

Your account: Access account settings, sign out, or switch users.

Power: Lock, sleep, shut down, or restart.

Accessing Content in the Start Menu

1. Click "Start"or press the "Windows" key. ***or***
2. Pinned apps are displayed in the Start Menu. **To see an alphabetical list of installed apps**, click the "All" button. All > A list of previously accessed files also appears in the Start Menu. **To view more files**, click the "More" button. More > **To return from either list**, click the "Back" button. < Back
3. **To search for apps or content**, click into the Search bar and begin typing. **To refine the search**, click the "Apps," "Documents," or "Web" buttons.

Accessing the Quick Link Menu

To access advanced tools for Power Users (Disk Management, Device Manager, etc.), *right-click* "Start." ***Or***, press the "Windows" and "X" keys. +X

Arranging Apps in the Start Menu

To move an app to a new location in the pinned list, click and drag it, then release. **To move an app to the top of the pinned list**, *right-click* the app and select "Move to front" from the menu that appears.

Pinning Apps to the Start Menu

To pin an app to the Start Menu:

1. *Right-click* on an app on the Desktop, within the Start Menu app list, or when viewing in File Explorer.
2. Select "Pin to Start" from the menu.

To remove an app from the pinned area:

1. *Right-click* the app.
2. Select "Unpin from Start" from the menu.

Customizing the Start Menu

Adjust settings to change what is displayed in the Start Menu ("Most Used" or "Recently Added" apps, etc.).

1. *Right-click* "Start."
2. Select "Settings" in the menu that appears.
3. Click "Personalization" on the left side of the window, then "Start" to the right.
4. Use sliders to toggle settings "On" or "Off." On

Pinning Apps to the Taskbar

1. *Right-click* an unpinned app in the Taskbar, on the Desktop, in the Start Menu, or in File Explorer.
2. In File Explorer or on the Desktop, click "Show more options." In the Start Menu "All" app list, roll over "More." Otherwise, skip to step #3.
3. Select "Pin to taskbar" from the menu.
4. **To rearrange pinned apps**, click and drag them within the Taskbar.

Using Jump Lists in the Taskbar

Jump Lists display an app's pinned files, recently used files, and related tasks from the Taskbar. **To display the Jump List for an app button in the Taskbar**, *right-click* it (or swipe up).

Customizing the Taskbar

1. Right-click on a *blank* spot in the Taskbar.
2. Click the "Taskbar settings" command.
3. **To change Taskbar icons**, click the sliders to toggle settings "On" (to show) or "Off" (to hide). On
   TASKBAR ITEMS: Add or remove items from the Taskbar ("Search," "Task view," "Widgets").
   SYSTEM TRAY ICONS: Show or hide icons for "Pen menu," "Touch keyboard," and "Virtual touchpad" in the right corner of the Taskbar.
   OTHER SYSTEM TRAY ICONS: Select icons to appear in the Taskbar corner and Hidden icon menu.
4. **To change Taskbar behaviors like alignment, auto-hide, and badges**, check (to turn on) or uncheck (to turn off) settings in the **TASKBAR BEHAVIORS** section.
5. Click the "Close" button in the upper-right corner of the window when finished.

Using the Notification Center

The Notification Center provides a unified place on the Taskbar for system notifications. It appears above the date/time area. The notification bell icon at the right end of the Taskbar lights up when there are new messages.

1. **To open the Notification Center**, click the Time and Date area at the right end of the Taskbar ***or*** swipe in from the right screen edge.
2. **To view the details of a message**, click it.
3. **To dismiss a message**, hover over it and click the "Clear" button.
4. **To remove all messages**, click the "Clear All" button. Clear all
5. **To choose which notifications to receive**, right-click anywhere within the Notification Center and select "Go to notification settings," make desired selections, then click the "Close" button.
6. **To manually turn on do not disturb and silence notifications**, click the bell icon with the "zZ" overlay in the Notification Center. **To turn do not disturb off**, click it again.
7. **To close the Notification Center**, click anywhere outside of the Notification Center.

Using Quick Settings

Quickly change common settings like audio, display, and connectivity without opening the "Settings" app.

1. Click one of the Hardware Status Indicators at the right end of the Taskbar.
2. Adjust settings as desired.
3. **To view additional quick settings**, click the scroll buttons on the right edge of the panel.
4. **To access the full settings for a quick setting action**, right-click the action and select "Go to Settings."
5. **To rearrange the actions within the panel**, click and drag them to the desired location.
6. **To open the full "Settings" app**, click its button.

Using and Customizing Widgets

Widgets are small snapshots that display dynamic content from apps and services like weather, news, and stocks, on the Windows desktop. They appear on the widgets pane, which can be customized to reflect your specific interests.

1. **To show the Widgets pane**, click or hover over the "Widgets" button at the far-left end of the Taskbar.
2. **To add more widgets to the Widgets pane**, click the "Add widgets" button near the top of the pane (below Insights), then click the "Pin" button next to the widget(s) to add. Pin When done, click the "Close" button in the "Pin widgets" window.
3. **To edit a widget's settings**, click its "More options" ellipsis button in the upper-right corner of the widget. **To change the widget's size**, select a size from the drop-down. **To make further changes**, select "Customize widget" from the menu, if available, and make selections. **To remove a widget**, select "Unpin widget" from the menu.
4. **To close the Widgets pane**, click the "Widgets" button in the taskbar again or move your mouse pointer away from it if you opened it by hovering.
5. **To remove the Widgets button from the Taskbar**, *right-click* in an empty spot on the Taskbar, select "Taskbar settings," then click the "Widgets" toggle button to turn it off.

TeachUcomp, Inc.

www.teachucomp.com
info@teachucomp.com
877.925.8080

File Explorer: Folders and Files

Understanding File Explorer

File Explorer is a window that allows you to explore and manage the files and folders contained within your computer and within your associated OneDrive account.

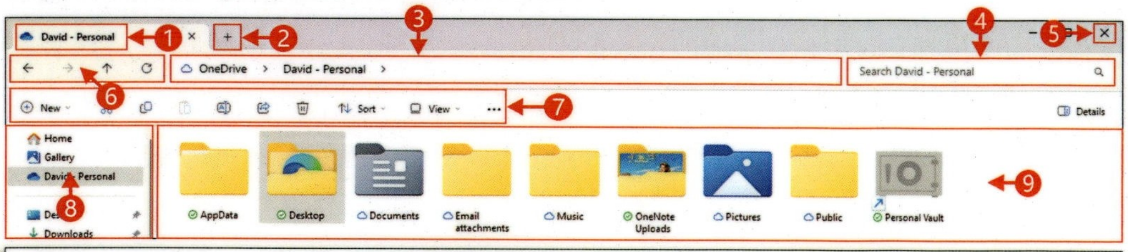

1. Tab Name
2. Add New Tab
3. Address Bar
4. Search Bar
5. Close Button
6. Navigation Buttons
7. Ribbon
8. Navigation Pane
9. Content Pane

Opening and Closing File Explorer

1. **To open File Explorer**, click the "File Explorer" icon in the Taskbar. ***Or***, press the "Windows" and "E" keys on your keyboard. ***or*** +E
2. **To close File Explorer**, click the "Close" button in the upper-right corner of the window.

Understanding the Ribbon

The **Ribbon** is displayed above the Navigation and Content panes and provides tools and options to work with files and folders. **To view additional options**, click the "See more" ellipsis button at the right end of the Ribbon. When an item is selected in the Content pane, the available buttons in the Ribbon that can be used on the selected item are activated.

Viewing Folders and Files

The Navigation pane on the left displays available folders and drives. The Content pane on the right displays the contents of the selected drive or folder. **To change what is viewed**, click the "View" button in the Ribbon.

1. **To preview content of certain files (images, videos, etc.)**, select "Preview pane" from the menu.
2. **To display certain file details**, such as size and creation date, select "Details pane" from the menu.
3. **To hide or show the Navigation pane**, select "Show| Navigation pane" from the menu.

To change display options in the Navigation pane:

1. Click the "See more" ellipsis button in the Ribbon and select "Options" from the menu.
2. Click the "View" tab in the "Folder Options" window.
3. Scroll down to "Navigation pane" in the "Advanced settings" section, and make selections:
   Always show availability status: Displays the current status of OneDrive files.
   Expand to open folder: Displays folders in the Navigation pane as they are opened at the right.
   Show all folders: Displays folders that are hidden by default (Control Panel, Recycle Bin, etc.).
   Show libraries: Displays libraries.
   Show Network: Displays local network devices.
   Show This PC: Displays contents of your computer.
4. Click "OK."

Navigating in File Explorer

1. **To view folder or drive content in the Content pane,** click its folder in the Navigation pane.
2. **To view subfolders**, click the right-pointing arrow to the left of the folder. **To hide (collapse) the subfolders**, click the downward-pointing arrow.
3. **To jump directly to a folder**, click its name in the address bar. **To jump to its subfolder**, click the right-pointing arrow to the right of the folder name and then select the subfolder.
4. **To navigate between viewed folders**, click the navigation buttons to the left of the address bar.
   Back/Forward
   Move up one level to its parent folder
   Refreshes content of window

Changing Folder Views

1. Click the "View" button in the Ribbon.
2. Select a view from the drop-down.

To add or remove columns in the "Details" view:

1. Select the "Details" view.
2. *Right-click* any column heading.
3. Check (show) or uncheck (hide) columns. For a full list or to change column positions, select "More" at the bottom, make selections, and then click "OK."

Customizing Home View

Home (or "Quick access") is the default view. It displays frequently or recently visited folders and files.

1. Click the "See more" ellipsis button in the Ribbon and select "Options" from the menu.
2. **To change the default view**, click the "Open File Explorer to" drop-down and select an option.
3. Make selections to determine how to open items and whether to display recently used files and frequently used folders.
4. Click "OK."

To add or remove Home (Quick access) items:

1. *Right-click* the item in the either pane.
2. Select "Pin to Quick access," "Unpin from Quick access," or "Remove from (*section name*)."

Sorting Folders and Files

Note: You cannot sort in Home (Quick access) view.

1. Click the "Sort" button in the Ribbon.
2. Select the attribute by which to sort. **To see additional options**, roll over "More."
3. **To change the sort order**, select "Ascending" or "Descending."
4. **To keep folders and files grouped together based on a specific criteria (date, type, etc.)**, select "Group by" and then make a selection or select "(None)" to remove any grouping already applied.

Opening Another Window or Tab

To open a new window, *right-click* a folder within File Explorer and select "Open in new window," *right-click* the "File Explorer" icon in the Taskbar and select "File Explorer," or press the "Windows" and "E" keys. +E
To open a new tab, *right-click* a folder within File Explorer and select "Open in new tab" or click the "Add new tab" button to the right of any open tab(s).

Creating a New Folder

1. Navigate to the location for the new folder.
2. Click the "New" button in the Ribbon.
3. Select "Folder" from the drop-down menu.
4. Type the folder name and press the "Enter" key.

Viewing Folder and File Properties

1. Select a folder or file.
2. Click the "See more" ellipsis button in the Ribbon and select "Properties" from the menu.
3. **To view various properties**, click the tabs.
4. Make any changes, as desired, and click "Apply."
5. Click "OK" when done.

Selecting Folders and Files

When an item is selected, it appears highlighted. Selected items have the focus of the Windows application and any commands you subsequently perform will be applied to the selected items only.

1. **To select a single item**, click it once.
2. **To select a group of items**, click and drag across the items starting from a blank part of the window.
3. **To select multiple adjacent items**, click the first item, press and hold the "Shift" key, then click the last item.
4. **To select multiple non-adjacent items**, hold down the "Ctrl" key while clicking the items.
5. **To select all items**, click the "See more" ellipsis button in the Ribbon and select "Select all" from the menu. Select all ***Or***, press the "Ctrl" and "A" keys.
6. **To select everything other than what is currently selected**, click the "See more" ellipsis button in the Ribbon and select "Invert selection." Invert selection
7. **To deselect an item**, press the "Ctrl" key and click the item.
8. **To deselect all items**, click into a blank area. ***Or***, click the "See more" ellipsis button in the Ribbon and select "Select none." Select none
9. **To select items using checkboxes**, enable the feature by clicking the "View" button in the Ribbon and selecting "Show," then "Item check boxes" from the menu. Hover over an item and click on the checkbox that appears to the left. Item check boxes

Renaming Folders and Files

1. Select the item to rename.
2. Click the "Rename" button in the Ribbon. ***Or***, press the "F2" key on your keyboard.
3. Type the new name and press the "Enter" key.

Deleting Folders and Files

Local items are sent to the Recycle Bin when deleted and can be restored later until permanently deleted. Network items are automatically permanently deleted.

1. Select the item(s) to delete.
2. **To delete the item(s)**, click the "Delete" button in the Ribbon ***or*** press the "Delete" key. **To bypass the Recycle Bin and permanently delete**, press the "Shift" and "Delete" keys instead.
3. Click "Yes" to confirm, if prompted.

Restoring Deleted Items

1. *Double-click* the Recycle Bin on the Desktop or select it within File Explorer.
2. Select the item(s) to restore.
3. **To return the item(s) to the original location(s)**, click the "Restore the selected items" button in the Ribbon. Restore the selected items ***Or***, drag the items to a different location.

Emptying the Recycle Bin

Emptying the Recycle Bin permanently deletes the items within it. Use caution when emptying.

1. *Double-click* the Recycle Bin on the Desktop or select it within File Explorer, then click the "Empty Recycle Bin" button in the Ribbon. Empty Recycle Bin ***Or***, *right-click* on the Recycle Bin on the Desktop or in File Explorer and choose "Empty Recycle Bin."
2. Click "Yes" to confirm.

Moving and Copying Folders & Files

First select the item(s) to move or copy, then:

1. **To move the item(s)**, drag and release the item(s) onto another folder, window, tab, the Navigation pane, or the Desktop.
2. **To copy the item(s)**, hold down the "Ctrl" key while dragging.

Or, using Cut, Copy & Paste:

1. **To copy the item(s)**, click the "Copy" button in the Ribbon. ***Or***, press the "Ctrl" and "C" keys. **To move the item(s)**, click the "Cut" button in the Ribbon. ***Or***, press the "Ctrl" and "X" keys.
2. Navigate to the new destination and click the "Paste" button in the Ribbon. ***Or***, press the "Ctrl" and "V" keys.

Creating Shortcuts

Shortcuts allow you to quickly open or access files, folders, or apps from different locations. Shortcuts appear with a small arrow in the icon's lower-left corner. Deleting, moving, or renaming a shortcut does not affect the original file.

1. *Right-click* the original item.
2. Select "Show more options" in the menu.
3. Select "Create shortcut" from the menu, then drag the shortcut to the new location. ***Or*, to quickly add a shortcut to the Desktop**, roll over "Send to" and select Desktop (create shortcut)" from the menu.

Searching in File Explorer

1. Navigate to a folder and click into the search bar to the right of the address bar.
2. Enter a search term. Matching file names, folder names, and content automatically appear in the Content pane as you type.
3. **To narrow the search only to certain locations**, click "Search options" button that appears in the Ribbon and select an option.
   All subfolders: Subfolders of the selected folder.
   Current folder: Only the selected folder.
4. **To refine the search further**, click the "Search options" button and make selections under "Date Modified," "Kind," and "Size" from the menu, and select (to include) "System files," "Zipped (compressed) folders," or "File contents" in the search. For faster searching, select "Change indexed locations," make changes, click "OK," then "Close."
5. Click "Close search" when done.

Zipping and Unzipping

Zip a file or folder into a widely supported archive format to compress its contents into a single "zipped" file that can be sent to others or simply compressed for storage. Zipped files appear as a folder with a zipper and act like any other folder.

To create a zipped file:

1. Select the item(s) to zip.
2. Click the "See more" ellipsis button in the Ribbon or *right-click* on one of the selected items, then select "Compress to ZIP file" from the menu.
3. Rename the zipped file, if desired, and press "Enter."

To unzip (extract) a file:

1. Select the zipped folder.
2. Click the "Extract all" button in the Ribbon.
3. Click the "Browse" button, navigate to a destination folder, and click the "Select Folder" button.
4. Click the "Extract" button.

To extract only specific files from a zipped file:

1. Open the zipped folder and select the item(s).
2. Move or copy the items as usual.

Using Libraries

Libraries let you manage files from multiple folders in one place, making it easier to find and manage them. Since files in libraries are linked, not copied, changes made to files in linked folders in a library also occur in their original locations. By default, Windows does not display libraries.

1. **To display libraries**, click the "See more" ellipsis button in the Ribbon, select "Options," click the "View" tab, scroll down to "Navigation pane," check "Show libraries," then click "OK."
2. **To view and manage libraries**, click the "Libraries" button in the Navigation pane.
3. **To create a new library**, click the "New" button in the Ribbon and select "Library."
4. **To rename a library**, *right-click* its name, choose "Rename," type the new name, and press "Enter."
5. **To add a folder link to a library**, *right-click* a folder in the Navigation pane, select "Show more options," then "Include in library," and select a library or "Create new library."
6. **To remove a folder**, select the library name in the Navigation pane, click the "See more" ellipsis button in the Ribbon, select "Properties," select the folder, click "Remove," and then "OK."

Apps

Understanding Apps

"App" is an abbreviation for "application." An app is a software program that runs through a web browser, or offline on your computer, tablet, smartphone, or other device. Apps for desktops and laptops are often called desktop apps, while those for mobile devices are called mobile apps. Windows apps are either included with Windows or can be downloaded from the Microsoft Store.

Searching for Apps

1. Open the Start Menu. ***Or***, click into the search bar in the Taskbar. ***Or***, press the "Windows" and "Q" keys on your keyboard. ⊞+Q
2. Begin typing the app name.
3. **To filter the search to show apps only**, click the "Apps" button after typing the search term.

Starting Apps

1. Click "Start" ***or*** press the "Windows" key.
2. **To start an app**, click it.

Or

On the Taskbar, click a pinned app. ***Or***, on the Desktop, double-click the app's shortcut.

Closing Apps

Click the app's "Close" button. ***Or***, press the "Alt" and "F4" keys on your keyboard. ***Or***, select "File| Close" or "File| Exit" from the app's menu bar or ribbon. ***Or***, hover over an app's icon in the Taskbar until a thumbnail appears, and click the thumbnail's "Close" button. ***Or***, *right-click* an app's icon in the Taskbar and choose "Close Window" or "Close all windows."

Minimizing Apps

Click the "Minimize" button in the upper-right corner of the app. ***Or***, press the "Windows" and the down arrow keys (twice if app is maximized). ⊞+▼

To minimize all open windows other than the current window, press the "Windows" and "Home" keys. **To minimize all open windows and show the Desktop**, press the "Windows" and "M" keys. ⊞+M

Maximizing and Restoring Apps

Click the "Maximize" button. ***Or***, press "Windows" and the up arrow keys. ⊞+▲ ***Or***, *double-click* its title bar. The "Maximize" button then changes to the "Restore" button. **To restore the app to its previous state**, click the "Restore" button, double-click its title bar, ***or*** press the "Windows" and Down-Arrow keys. ⊞+▼

Resizing Apps

1. Hover over the edge of a non-maximized window until the mouse pointer turns to a double-pointed arrow.
2. Click and drag the edge of the window.

Moving Apps

Click on an empty part of an app's title bar and drag it to its new location.

Opening App Files

By default, when a file is opened outside of an application, Windows will attempt to open it in its native application, if it exists on your computer (e.g. opening a Word document will launch Microsoft Word). For some files that can be opened by multiple applications, you will be prompted to choose one to open the file.

1. Navigate to the file on your computer (see "Navigating in File Explorer").
2. Double-click the file.
3. If prompted with "Select an app to open this (*file extension*) file," select an app from the list, look for an app in the Store, or click the "Choose an app on your PC" link to search further. **To make your selection the default to open the file type**, click the "Always" button, otherwise click "Just once."
4. Click "OK."

Saving App Files

To overwrite a file that has already been saved, click the "File" tab in the Ribbon or menu bar of the app and click the "Save" command. ***Or***, press the "Ctrl" and "S" keys.

To save a file for the first time or to save with a different name or location:

1. Click the "File" tab in the Ribbon or menu bar of the app and click the "Save As" command.
2. Click to select a location ("This PC," "OneDrive," or other location, such as a SharePoint Server for some Microsoft 365 subscribers).
3. Click a folder name or the "Browse" button.
4. Navigate to the location where you want to save the file in the "Save As" dialog box.
5. Type a name in the "File name" text box and click the "Save" button.

Switching Between Apps

Icons appear in the Taskbar for open apps, apps that have been pinned to the Taskbar, or that require attention. Icons change depending on their status.

- **Closed:** Solid background.
- **Open:** Short line at bottom of icon.
- **Current Window:** Light background, long line.
- **Multiple Windows:** Icon appears as a stack.
- **Needs Attention:** Flashing icon.

To switch to another app, click its icon in the Taskbar. For an app with multiple open windows, select from the thumbnails that appear when the icon is clicked. **To preview those open windows**, hover over the thumbnails. **To cycle through open apps**, press the "Alt" and "Tab" keys repeatedly.

Switching Apps Using Task View

Use Task View to quickly view all open apps and virtual desktops and easily switch between them.

1. Click the "Task View" icon in the Taskbar ***or*** press the "Windows" and "Tab" keys. ⊞+Tab
2. Click an app.

App Snapping

App Snapping allows you to lock up to four open app windows to the sides of your screen for easier viewing.

1. On an open app, press the "Windows" and either the right or left arrow key to snap it to that side. ***Or***, click and drag the app's title bar to the left or right edge of the screen and release when the mouse pointer reaches the edge. ⊞+◀or▶
2. **To then snap the app into the upper or lower quadrant of the screen**, *press the "Windows" key again* and press either the up or down arrow key. ⊞+▲or▼
3. Windows displays other open apps in an empty quadrant. **To snap an app to that quadrant**, click it.

To snap apps using Snap Layouts:

1. Hover over the "Maximize" or "Restore" button of an open app. A selection of Snap Layouts appears.
2. Click a layout to select it. The current app snaps to the first open area.
3. Windows displays other open apps in an empty area. **To snap an app to that area**, click it.

Using Virtual Desktops

Create multiple, separate desktops that can each display different open windows and apps.

1. Click the "Task View" button in the Taskbar.
2. Click the "New desktop" button.
3. Click the new desktop's thumbnail, then open the desired apps. Each desktop's Taskbar displays its own set of icons for any open apps.
4. **To switch between desktops**, click the "Task View" button and click the desired desktop's thumbnail.
5. **To delete a desktop**, click the "Task View" button, hover over the thumbnail, then click "Close."
6. **To move an open app to a different Desktop**, click the "Task View" button, *right-click* an app window, click "Move to," then select the destination.

Settings

Using the Settings App

The Settings app provides a centralized location to access and manage system settings, devices, users, personalization, internet access, security, and more.

1. Click "Start," then the "Settings" button. ***Or***, *right-click* "Start," then select "Settings" from the menu. ***Or***, press the "Windows" and "I" keys on your keyboard. + I
2. Click a setting group ("System," "Bluetooth & devices," etc.) from the list on the left, then select a setting to the right.
3. Make desired changes using the various buttons, drop-downs, and sliders. Changes are automatically saved. Some settings have "Related settings" below them that you can click to make additional changes. Make the changes in any dialog boxes that appear and click "OK" to save, when necessary.
4. **To return to the previous Settings screen**, click the "Back" button in the upper-left corner.

Using the Control Panel

The Control Panel provides another way to access various system and component settings.

1. Click "Start," then type "control panel."
2. Click "Control Panel" when it appears.
3. Click a Control Panel option. ***Or***, if "View by" is set to "Category," click a category link to see options.
4. Make the changes in any dialog boxes that appear and click "OK" to save, when necessary.

Customizing the Desktop and Lock Screen Backgrounds

1. Open the Settings app and click "Personalization."
2. Click "Background" or "Lock screen" to the right.
3. **To use a solid color (for Desktop)**, select "Solid color" from the "Personalize your background" drop-down and select a color. ***Or,*** for a custom color, click the "View colors" button, click into the color field, move the slider, and click "Done."
4. **To use an image**, select "Picture" from the drop-down. Select an available picture. ***Or***, click "Browse photos," navigate to an image and select it, then click the "Choose picture" button.
5. **To use a slideshow**, select "Slideshow" from the drop-down. Select an available album. ***Or***, click the "Browse" button, navigate to a folder of images, select it, then click the "Choose this folder" button.
6. Make any available additional setting selections.

Managing Apps

1. Open the Settings app and click "Apps" at the left.
2. Click "Installed apps" to the right.
3. **To managed installed apps**, click the "More options" button at the right end of the app in the list and select "Modify" or "Uninstall" and follow the prompts.

Setting Up Wi-Fi

1. Open the Settings app and click "Network & internet" at the left.
2. Select "Wi-Fi" from the menu to the right.
3. Click "Show available networks" in the list below.
4. Click on an available network and click "Connect."
5. Enter a password and click "Next," if necessary.
6. Follow any additional prompts to complete setup.

Managing Wi-Fi and Airplane Mode from the Quick Settings Panel

1. Click one of the hardware status indicators at the right end of the Taskbar.
2. **To toggle all wireless communication on or off**, click the "Airplane mode" button.
3. **To manage Wi-Fi**, click the arrow next to the Wi-Fi symbol. **To select a network**, click it, then click "Connect." **To toggle Wi-Fi on/off**, click its button at the top.

Troubleshooting and Help

Shutting Down a Frozen App

1. *Right-click* on "Start" and select "Task Manager." ***Or***, press the "Ctrl"+"Shift"+"Esc" keys.
2. Click the "Processes" tab and locate the app.
3. If the app's status is "Not Responding," select the app in the list.
4. Click the "End Task" button. End task

Troubleshooting Print Jobs

Check that status of print jobs and pause, cancel, or restart jobs as needed.

1. Open the Settings app, click "Bluetooth & devices" at the left, "Printers & scanners" to the right, select the printer in the list, then click "Open print queue" under "Printer settings." ***Or,*** *double-click* the "Printer" icon at the right end of the Taskbar.
2. In the "Print Queue" dialog box, *right-click* the print job in the list.
3. **To change the print job status**, select "Pause," "Resume," "Restart," or "Cancel." Click "Yes" to confirm, if prompted. Note: If you select "Cancel," some pages may finish printing that were stored in the printer's memory buffer. ***To stop printing***, try turning the printer off and then turning it back on after a moment or two to clear the memory buffer.

Getting Help with the Get Help App

The Get Help app provides self-help with instant answers, troubleshooting, support articles, and more.

1. Open the Get Help app. (See "Searching for Apps" and "Starting Apps").
2. Type a question. Matches appear as you type.
3. Click a result to open it and finish searching. ***Or***, continue typing and press "Enter" to show matches.
4. When done, click the "Close" button.

Users and Sign-In Options

Adding Users

Open the Settings app and click "Accounts," then:

To add a family user:

1. Select "Family" to the right.
2. Click the "Add someone" button. Add someone
3. If adding an adult family member, enter the new user's email address, click "Next," select a role, and click the "Invite" button. If adding a child, click the "Create one for a child" link instead, then enter the email address and click "Next." Enter any required information, clicking "Next" after each step.
4. Click "Close" when done. An invitation is automatically sent to the new user's email address.

To add other users (work or school):

1. Select "Other users" to the right.
2. Click the "Add account" button. Add account
3. **To add a user with a Microsoft account**, enter their email address or phone number. **To create a local account**, click the "I don't have this person's sign-in information" link and follow the prompts.
4. Follow the prompts to complete account setup, including setting a password or PIN.
5. Click "Close" once the account is created.

To connect a personal Windows device to an organization's network and resources:

1. Select "Access work or school" to the right.
2. Click the "Connect" button. Connect
3. Follow the on-screen prompts to sign in with work or school credentials.
4. If prompted, choose to join the device to the organization's Microsoft Entra ID.
5. Follow the remaining steps to complete the setup.

Switching Users

On the sign-in screen, click a different user name in the bottom-left corner. Sign-in as usual. ***Or, to switch to another user from the Desktop***, click "Start" and then the "Account" button, choose a different user, then sign-in as usual.

Configuring Sign-In

1. Open the Settings app and click "Accounts."
2. Click "Sign-in options" in the menu at the right.
3. Click "Change," "Change PIN," "Add," or "Remove" under "Password," or "PIN." **To set up sign-in with a fingerprint, facial recognition, or security key** (requires compatible hardware), click "Set up" or "Manage" under the desired sign-in option.
4. Enter password/PIN if prompted. Make changes: **Password**: Type current, then new password twice. **PIN**: Type current, then new PIN twice.
5. Click to confirm changes when prompted.

Keyboard Shortcuts

General Shortcuts

Select All	**CTRL + A**
Copy	**CTRL + C**
Cut	**CTRL + X**
Paste	**CTRL + V**
Delete	**CTRL + D**
Undo	**CTRL + Z**
Redo	**CTRL + Y**
Find	**CTRL + F**
Save	**CTRL + S**
Open	**CTRL + O**
Print	**CTRL + P**
Switch Between Windows	**ALT + TAB**
Close Active Document	**CTRL + F4**
Shut Down App	**ALT + F4**
Sign-Out/Task Manager Options	**CTRL + ALT + DELETE**
Open Task Manager	**CTRL + SHIFT + ESC**

Windows Logo Key Shortcuts

Open/Close Start Menu	
Open File Explorer	**+ E**
Open Quick Settings	**+ A**
Open Settings App	**+ I**
Show Desktop	**+ D**
Lock Screen	**+ L**
Minimize Active Window	**+ Down Arrow**
Maximize Active Window	**+ Up Arrow**
Minimize All Windows	**+ M**
Minimize All But Active Window	**+ HOME**
Restore All	**+ SHIFT + M**
Snap App Left	**+ Left Arrow**
Snap App Right	**+ Right Arrow**
Zoom In	**+ Plus Sign (+)**
Zoom Out	**+ Minus Sign (-)**
Exit Magnifier/Reset Zoom	**+ ESC**

Touch Actions

Using a Touch Interface

Click	**Tap**
Double-Click	**Double-Tap**
Right-Click	**Touch and hold for box & release**
Scroll	**Drag or Swipe**
Pinch/Stretch	**Move two fingers towards each other or away to shrink or enlarge**
Rotate	**With 2 or more fingers, turn hand**

ISBN 978-1-958446-02-7

Made in the USA